Nothing You

See Is Real

Table of Contents

Our Physical Universe Is Merely An Illusion

Astronomy Domine

Evolution

Proof and Historical Evidence

Enuma Elish (Epic of Creation)

Science Facts

Ancient Astrology

Billy Meier Befriends Aliens

Dawning of a New Age

Laura Eisenhower and Andrew Basiago

Teleportation and Levitation

Quantum Teleportation

Holograms

Copyright © 2022

This book is sold subject to the condition that it shall not, by way of trade or otherwise, be lent, resold, hired out, or otherwise circulated without the publisher's prior consent in any form of binding or cover other than that in which it is published and without a similar condition including this condition being imposed on the subsequent publisher.

Our Physical Universe Is Merely An Illusion

At the start of all my books I pay homage to Mauro Biglino. Signore Biglino is an important minister of the catholic church. He is the inspiration for the whole world. I heartily recommend the awesome work of Mauro to all my readers.

Throughout history we have been instructed by self-styled 'experts'. This has been especially apparent in all modern religions and formal education. In reality those who instruct were themselves spoonfed information they believed to be correct. Why did they believe it to be true? Because they were told by their peers that it *was* the truth. This passage of apparent facts from one generation to another has resulted in a picture of the world and especially history that is far from accurate. In fact deliberately instilled disinformation has placed the people of the world into a profound state of ignorance.

If I told you that *all* we have been taught throughout history is largely fabrication would you accept my statement at face value? I do hope not. However you have accepted at face value *all* that your peers have taught you. Their policy has always been to provide information that *best suits their needs* and not your own.

A perfect example is the Scriptures. Genesis was written by Hebrews who themselves were captives in Babylon. Wishing for an end to enslavement they scoured the Library of Ashurbanipal. The two Hebrew authors of Genesis attempted to explain why their gods had abandoned them. When they found passages in the **Enuma Elish** that explained the origins of our world and the gods' creation of mankind they fabricated a story of Satan (the Serpent) leading the first couple astray. Owing to Eve's transgression in disobeying her gods all mankind (in particular all females) were doomed to a life of pain and torment. Since then females have suffered a 'bad press'. This act of eating an apple from the Tree is entirely fiction. Simply by reading the story in Hebrew you will see that it makes no sense to all.

In this book I intend to lay out facts that when compounded will positively influence *your* decision to accept that the reality we believe we exist in is in fact an illusion.

We shall begin by unravelling all the 'facts' that you have hitherto been informed are absolutely the truth. As we progress I may offend some but surely *the* Truth is the fundamentally most important aspect and must be pursued no matter what.

This book is written after nearly sixty years of research. I have sought the elusive Utopia, happiness, peace and contentment and explored spirituality. I have searched religions, science and mysticism for the answer to the fundamental questions:

- Who am I?
- Why am I here?

I searched for the answers in formal education, formal religious practices and the sciences as well as in deep meditation. I believe I may have discovered the Truth. It is so simple yet fantastic only a handful of human beings discover **It**. A handful from billions.

Until the fifteenth century we were taught that the Earth was flat. Then we were taught that the sun revolved around the

Earth. Humans are all made in the image of God and thus we rule planet Earth. There are no other lifeforms on the billions of planets in the Universe.

Yet recently cosmologists have informed us there are inhabitable planets by the bucket load and *our sun even has a twin*.

Astronomy Domine

More giant observatories and high powered telescopes have been built (both on the Earth and out in space) in the past twenty years than ever before in history. What are they focused on?

Many are tracking the passage of a new arrival to our solar system – the star system collectively referred to as Nibiru. Yes indeed our sun has a twin star. Full details later.

Why are constellations so important? Because the *very energies that affect us* change with the introduction of each new

Age. According to the Maya each Age lasts approximately 2250 years. There are twelve signs of the Zodiac so to return to a starting point on the Wheel of Life takes 2250 x 12 = 27000 years approximately.

This is not some mathematical exercise. Our entire solar system along with its twin system (Nibiru) orbits the Orion system and in turn this entire system orbits the Sirius system. In combination these systems skirt *the entire* Milky Way once every 27000 years. The ancients knew this and proceeded to construct stone observatories throughout the planet.

Our observatories may be a little more high tech but the process of tracking the precession of the Eqinoxes is identical. Advanced ancient civilizations called this 27000 year period the Great Year. It is referenced by the Maya, Atlanteans and in the Sanskrit scriptures. I suggest here you pick up my extremely popular series **Exposed History of Planet Earth.**

Recently we moved from the Age of Pisces (236 BC – 2014) to the Age of

Aquarius. Our planet has entered 'the Spring Awakening'. The position of our planet in space at this time, relative to the massive black hole at the centre of our Galaxy, places us in an unique position.

This 'Spring time' allows the planet itself and its inhabitants to accept transmissions from the black hole that are allowing us *to break free from this mundanity*, this illusion we think is reality. Yet there are those who wish us to remain imprisoned forevermore. Some say this is Satan, some say Illuminati some say Lucifer and some blame reptilians. It really does not matter who is to blame. For a full account of the Illuminati read my book **The Illuminati.**

The most important fundamental truth here is that we have the opportunity to exit the Cosmic Computer Game we have been engaged in for millenia.

Spreading love and compassion, tolerance and understanding and envisaging a perfect higher realm will cause you to break free from this illusion. Be like Alice and denounce the horrors of this world as figments of an overactive imagination. The choice is yours and I suggest you find an approach that best suits you. For me meditation on the Merkaba is perfect.

I repeat if you focus on worldly matters and bodily desires at this auspicious time you will once again become embroiled in the illusion for a minimum period of 27000 more years.

Meditation, following those who have love in their hearts, imagining and creating your own unique happiness and helping others to do the same goes a long way towards finding personal freedom. Forget what others are saying, trust yourself, your heart, your instincts. This is the time when Ascension is possible. As Jesus Christ, Enoch, Adam and Muhammad ascended now is the time for all good souls to return to their true home.

Evolution

Anthropologists and scientists have created a heartwarming explanation for the evolution and origin of the human race. Scientists in their zeal to postulate how the Earth itself came into existence have merely guessed at its origin. They have ignored the overwhelming evidence that proves humans are an engineered species that were from another planet many light years away. Their guesses have become accepted belief at least in the 'developed' countries. Yet 'primitive' peoples (despite recent indoctrination from the Western countries) have always known the Truth.

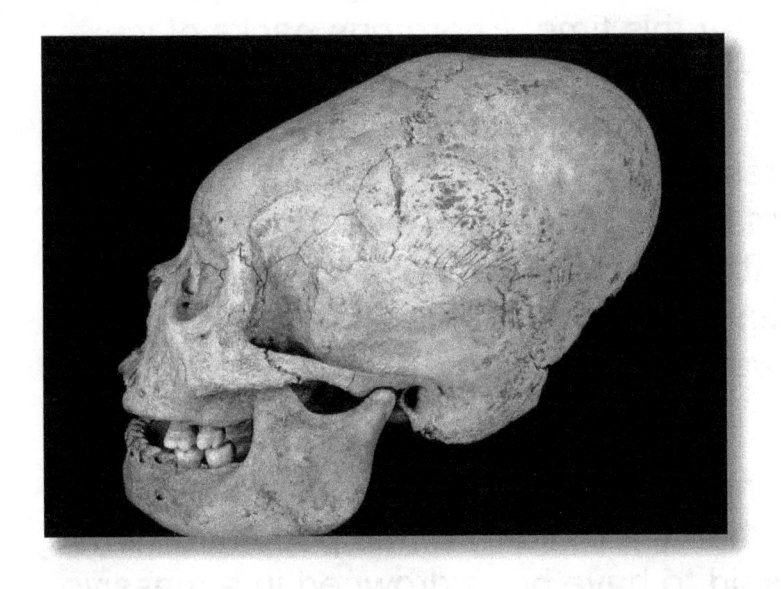

The book of Genesis tells us "There were giants in those days. They were called the Nephilim. They came from heaven to Earth. They took for themselves human wives"

In those days is an indeterminate period. Genesis was compiled around 250 BC so 'in those days' must be some time BEFORE 250 BC. When *was* this? Human history (at least written) may at the very least be traced to a Great Calamity that occurred perhaps 12,600 years ago. Scientists confirm that there was a 'sudden' warming of the planet after this time. Enormous packs of ice (miles deep across North America and Europe) melted at an alarming rate. The resultant rise in sea levels caused massive Tsunami to sweep across the Atlantic and beyond. Some believe at this time the wonderful continent of Atlantis was swallowed by the ocean.

The Great Flood is a period in antiquity when the majority of the population was said to have been drowned in a massive Tsunami that either swept the entire planet or at the very least Mesopotamia. Certainly there are Flood Myths throughout the world. In all these accounts a remnant of the human race was rescued by 'god'. Were there other survivors elsewhere?

Many writers (especially Zecharia Sitchin) specify that our ancestors were taught many things by E.B.E.s (Extraterrestrial Biological Entities). He firmly states that our ancestors were altered genetically to suit the purposes of these E.B.E.s. He believes (as I do in my series **Exposed History of Planet Earth** that we became enslaved by this race of giants.

Around 500,000 years ago the Neanderthal mysteriously vanished and 'modern' man 'appeared on the scene'. Archaeologists (adhering to Darwinism) and scientists have spent many fruitless years searching for the missing link between man and ape. There is *no* missing link. It was at this time that the Anunnaki genetically altered the Lulu of Eastern Africa to increase their intellectual capacity and thus serve *as useful slaves*. They became a slave race for the 'gods'. Mr. Sitchin and myself have written extensively on this subject and I shall not be repeating his work here. My first book on this subject is entitled **Arrival of the Anunnaki. I do however venture far further than Mr. Sitchin. I prove almost all our history is fake.**

On the other hand accepted doctrine is that 'somehow' a fantastic civilisation arose in the Middle East ca 3000 B.C. These were the Sumerians and they built Ziggurat and aqueducts and enjoyed the pleasures of bathing, sanitation and agriculture. They had a firm grasp of

Mathematics, measurement of time and astronomy. Pretty amazing for a bunch of savages.

Remains of the giants that walked the Earth (our Lords, our gods) are found *throughout the world*. Stepped pyramids, the Nazca lines, stone circles and enormous ziggurat are on display throughout the planet. Unfortunately the

Smithsonian Institute steals giants' skeletons.

Ironically the same archaeologists and scientists that 'buy into' Darwinism have ignored the giant skeletons they continually unearth. It is these skeletons that present the clue to why the missing link between Neanderthal and Homo sapien sapien *shall never be found*.

There *is no* missing link. Some 500,000 years ago some humans (in particular homo Africanus) were genetically altered to suit the purposes of our masters – the giants known to Sumerians as the

ANUNNAKI

Mama with Lulu

The Anunnaki are a race of humanoid giants that ruled over planet Earth for hundreds of thousands of years. Those they left in charge when they returned to their own star system have kept us all in ignorance ever since ca 1054 BC. They have been incredibly successful at spinning *a false history story* in order to maintain control of both religion and Government. Many are in fact descendants of hybridized Anunnaki/humans known as demigods.

The star system referred to by the ancient Sumerians as Nibiru has recently reentered our solar system as indeed it does *every 3600 years*. It comprises a brown dwarf star and three orbiting planets. Check out the NASA website and you tube to verify these statements.

Many Ooparts. Search the Web Please.

Our 'gods' travelled freely from Mars to the Moon to Earth and manned orbiting space stations. This is not so far fetched nowadays as we have achieved all the above ourselves relatively recently. Yes we have colonists on Mars. See my book

Of course we are taught *none of this* at school and mainstream science maintains silence on the subject of alien rulers in the dim past. Yet the evidence has piled so high that much of it lies gathering dust in Museums throughout the planet.

NASA footage is seriously airbrushed daily to hide the wonderful ancient cities of crystal and modern structures on the Moon. The pyramids and mausoleum on Mars are passed off as 'natural oddities'.

Perhaps this is all too much and you wish to return to accepted doctrine. If so close this book quietly. Was the Earth a 'chunk' of a star or a 'ball of swirling gases' that somehow solidified then cooled *over billions of years*? Did Darwin really discover the truth of evolution?

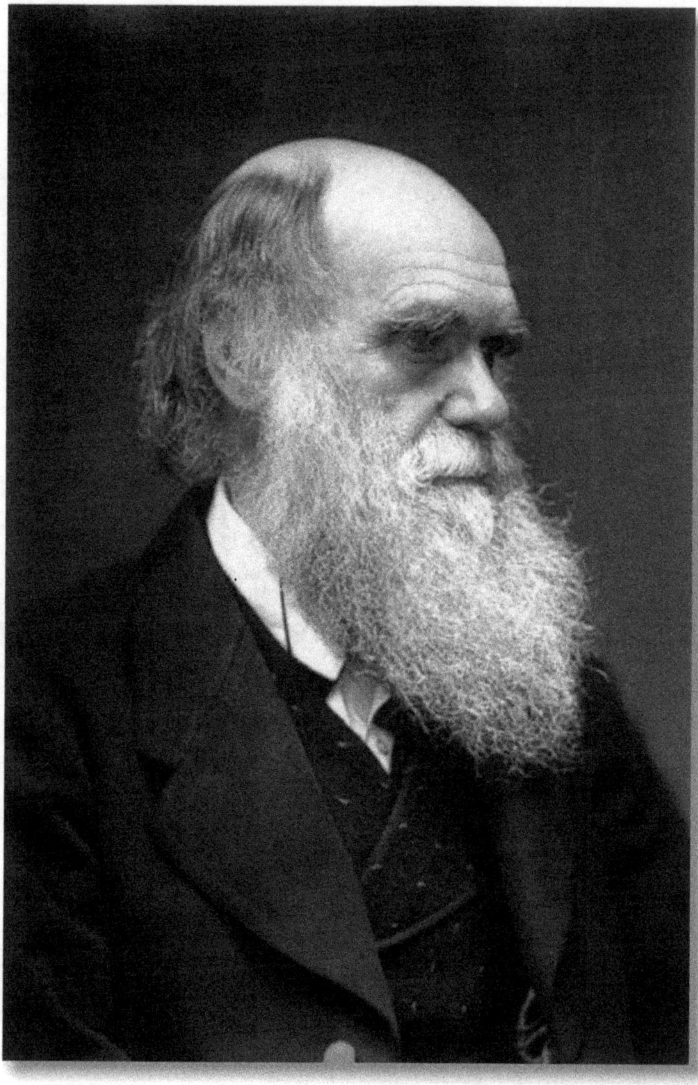

Darwinism is a rather lazy way of explaining how all life evolved from *a kind of Primeval Soup*. In his book The Origin of Species Darwin states that humans evolved from apes. If this theory is correct then *we should be able to mate with apes*. In fact we should still *be* apes. Apes are stronger than humans and obviously well adapted to their environment (if not, according to Darwin, *they would have evolved into something else*).

If apes evolved into humans then why are there still apes throughout the world and what conditions were present for apes to suddenly begin wearing high heels?

If the ocean is the ocean why are there particularly large whales and also tiny fish? Wouldn't all oceanic life evolve into the most advantageous form? I have lots more to say on this subject but I believe Darwin (despite his thousands of findings) didn't quite 'hit the nail on the head'. For a Victorian populous, enslaved by religion for thousands of years, a 'scientific' explanation was welcomed. At this time in history the general populous had had enough of religion being rammed down their throats, observing the Church becoming ever richer.

Scientists hypothesise that all life originated in the oceans. Since this theory became popular many have attempted to recreate life in the laboratory using electricity to simulate lightning. Now we all know that lightning (like electricity) is earthed by the ground itself. It also sends massive electrical charge *out* of our atmosphere. How is lightning able to super-excite chemicals in the oceans? How on Earth did all these amino acids gather together to produce a fish let alone a human? Somebody please explain. My head is sore with scratching.

No wonder scientists have not been able to duplicate *complex* life forms. They should be attempting to duplicate them *in Outer Space* where the electrical charge is far greater than in the Primeval Soupy oceans.

If Darwin was correct then why do we have more DNA in common with pigs than we do with apes? In fact we have much in common with the puffer fish. Nobody has told me we evolved from puffer fish. Incidentally 50% of human DNA is in common with a banana.

Because the period of evolution is over millions of years *any hypothesis* may be considered valid. I learned at school that owing to the drying of pools of water over thousands of years gills became lungs so 'fish' adapted to breathe air rather than water. Well when I observe pools they dry up pretty darn quickly. This evaporation would be far quicker millions of years ago when the Earth was much hotter. What were the silly fish doing in pools in the first place when there is the ocean to swim in?

Give or take a few more million years the 'air breathers' sprouted legs and took to the trees in fear of their lives. Some magically developed wings and others long tails to hold their credit cards.

The concept of escaping from dinosaurs by developing wings makes no sense at all as many species of dinosaur were also aviators. The mere thought of hiding from a T Rex in the branches of a tree *again* is preposterous as the hungry carnivore will simply uproot the tree or pluck the 'monkey' from its perch.

When the danger from dinosaurs was past (fortunately there was a cataclysm in the form of a giant comet or comets that wiped out all dinosaurs – phew just in time) the shy tree dwellers returned to the ground to begin foraging. They must have been extremely hungry over the previous few million years.

When the dinosaurs had ceased to exist (all dinosaurs, all at once, all around the world) there would be scant reason to take to the air to escape them. Then why didn't birds lose their wings and develop fingers, all the better to eat their KFC. After all we developed fingers whilst we were losing our pesky tails.

Much later clever 'apes' accidentally discovered the use of fire and that is why we have an abundance of internationally renowned chefs.

On a serious note I pose a question for you and scientists to answer. Given that humankind arrived on the world stage relatively recently they have since proceeded to dominate the planet in an extremely 'short period of time'. Why then did humans not evolve earlier? After all if natural selection is all about the fittest and finest adapting to the environment haven't we proven that we are by far the best? Or have we?

I make no apologies for poking fun at Darwinism. Charles himself refuted his findings several months before his death. He had primarily focused on the Galapagos Islands before presenting his theory to the Linnaean Society in 1858. He studied extensively the slow moving giant tortoises. My question is why did these tortoises not become dolphins or whales? The latter are far better adapted to their environment. If Darwin is correct there would be less than a hundred species on our beautiful planet not hundreds of thousands.

Humans have not physically adapted at all well to their environment. When the First People of North America and the Inuit were encountered by 'white man' the 'savages' died in their thousands of innocuous ailments such as the common cold. Actually the white guy deliberately introduced Smallpox into the indigenous population. That was men, women and children deliberately infected. Just a simple question if you pardon me for a second. Wherefrom did Coronavirus emerge? Strongest human *cannot* uproot a large tree as elephants can. Indeed as water is far more prolific than land why are we not living in the oceans?

We need a more realistic theory for how a species that is at the same time brown, black, yellow and pink-skinned could become the apparent Overlords of planet Earth. We must logically follow a regression back to our beginnings rather than hypothesising from which chemical bath we were sired.

Proof and Historical Evidence

Documentary evidence for the existence of the human race extends backwards to around the year 2600 BC. Sumerians and Egyptians were writing on baked clay tablets at this time. This only proves that there was *already* a language in play for both of these civilisations. The spoken word must have preceded the written word. Perhaps grunt grunt grunt.

Therefore we must begin our quest for the origin of our species by asking ancient Egyptians and Sumerians. Did *they record their history*? Are they able to present to *us* clues that indicate how the world and humankind were created? Indeed they are.

Sometime during his reign King
Hammurabi of Babylon (1792 BC to 1750

BC) captured the

inhabitants of the land we know today as the sorry State of Israel. King Hammurabi took prisoner thousands of Hebrews (there was no Judaism in those days).

After the Battle of Carchemish in 605 B.C. King Nebuchadnezzar again enslaved people from this region. They were once again held captive in Babylon.

It was during the latter captivity in Babylon that Genesis was written. It was written by two or more scholars, both in captivity and both with access to ancient knowledge.

Many Hebrews were allowed to return to their homeland around 538 BC (following the Persian overthrow of Babylon). The remainder of the Torah was completed thereafter.

I shall point out right away that the book of Genesis should not be taken as

Gospel. It was poorly copied from information contained in the Library of Ashurbanipal in Babylon. The book of Genesis (written by two or more authors) was a rough translation of the Enuma Elish (written three thousand years earlier) in Sumeria.

Enuma Elish (Epic of Creation)

When in the height heaven was not named and the earth beneath did not yet bear a name...

The primeval Apsu (who begat them) - and chaos (Tiamut, the mother of them both).
Their waters were mingled together and no field was formed, no marsh was to be seen.
When of the gods none had been called into being, none bore a name and no destinies were ordained.....

Then were created the gods in the midst of heaven - Lahmu and Lahamu were called into being....

Ages increased then Ansar and Kisar were created....

Long were the days then there came forth Anu.

These are the first two paragraphs of the Epic of Creation.

So Sumerians not only profess to have knowledge of a time before the Earth was created they also name the earliest 'gods'.

Compare this with Genesis:
In the beginning God created the heavens and the earth.... "Let there be light,"

No real clue as to how the Earth was created, just that it *was* created (which is self evident).

The Torah also originally spoke of *Gods*. It wasn't until 'Moses' (himself a deposed

Egyptian Pharaoh) led the newly named Israelites out of Egypt (every one of them a disgruntled Egyptian) that Gods became one God. The 'believers' were renamed Israelites. They took the name of Esau's brother (Jacob became Israel). However Israel also means follower of the Lord (god) Ra. Interestingly Ra is an *Egyptian* god.

At that time Hebrews still worshipped a pantheon of gods. In fact many Israelites returned to their 'false worship' after the death of 'Moses'. Consequently when Genesis was written there was still a belief in a pantheon of gods. They included:
Baal, Asherah, Yahweh, Astarte, Beelzebub, Amun, Nergal, Tammuz, Nabu, Bel

Many of these gods were shared with Assyrians, Sumerians and Egyptians.

It stands to reason if the Sumerian civilisation was the first then later civilisations would adopt their customs and their gods.

I repeat, all those who travelled to Israel 'on a forty year sojourn' were Egyptians as indeed was 'Moses'. What is most disconcerting is that Moses was not his real name.

The term Moses actually means saved from water. After many translations of the Torah into so many languages the 'child saved from drowning' later became known as Moses.

I shall briefly explain this important stage in history as I have covered it extensively in other books. 'Moses' was the son of a princess. Shortly after he was born the Vizier dreamt that the child (upon reaching manhood) would present a serious problem to the future monarchy. The Vizier strongly advised having the baby put to death. In fear of her child's life the princess spirited the baby away to a safe location. He was raised by foster parents. Bulrushes my arm.

In the ensuing years Egypt went through a dreadful period of turmoil as first one 'house' then another made claim to the throne. Not only civil war but war with other nations reduced the power of the empire. Eventually his mother became Queen Tiye.

Striking Similarity

After the death of his father (Amenhotep III) the 'child' was placed on the throne for a period of time. 'Moses' was crowned Amenhotep IV. His Queen was Neferteri. Whereas her tomb has been discovered his has not and never will be, because he died in Canaan or the USA.

The pair ruled Egypt for about 17 years from 1350 BC.

On day 13 month 8, in the fifth year of his reign, he insisted on becoming known as Akhenaten (devoted to Aten). Aten is the disk of the Sun (Sun God).

From the outset Akenaten became rather unpopular as he insisted upon worship of *one* God. Zoroastrianism was the original monotheistic religion (originating in modern day Iran). His foster parents were Zoroastrians. They had reared 'Moses' in their own belief system.

Destroying effigies of all the Gods and replacing them with a smiling sun face did not go down well with the majority of Egyptians.

Amenhotep IV clung onto his throne for a scant few years. Civil war again ensued and 'Moses' was banished to a satellite of the Egyptian empire known as Canaan. Coincidentally this is where modern day sorry state of Israel stands. He took with him like-minded Egyptians. This of course means the original Israelites were Egyptians. There were no Israelites until banished monotheists created the tiny state of Israel. Israel has been attacked on numerous occasions. In fact the whole of Canaan has 'changed hands' many times.

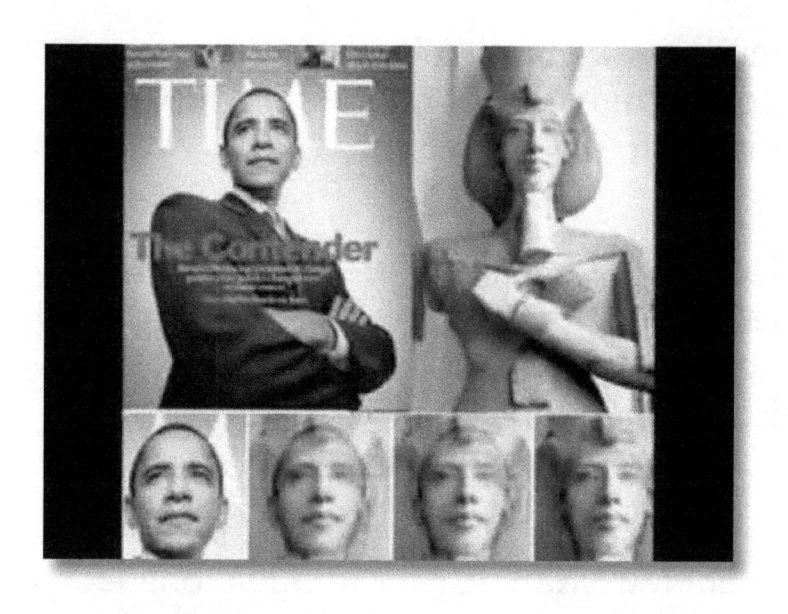

I have explained very briefly the time before the Exodus for a very important reason. Please prove me wrong (and many other authors who agree with my findings). If indeed I am correct then at least one segment of the Old Testament is incomplete. How much more?

Returning to the origins of our planet

Sumerian writing tells us that the first gods were named Lahmu and Lahamu. Much later a god named Anu enters the scene. We shall return to this extremely important god later.

How was Earth created?

According to the Enuma Elish:

He (Marduk) made a net to enclose the inward parts of Tiamat.
The four winds he stationed so that nothing of her might escape.
He created the evil wind, and the tempest, and the hurricane.
He sent forth the winds which he had created, the seven of them.
To disturb the inward parts of Tiamat, they followed after him.

I have skipped a great deal of the Enuma Elish (please read **Arrival of the Anunnaki** if you want the full script).

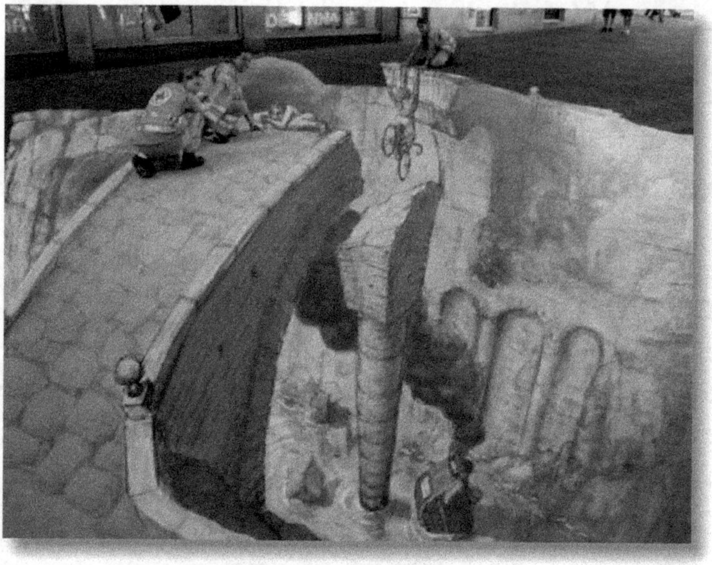

Marduk is a God

When Marduk was granted rulership of our planet, for a very 'short time' during the Ascension of Aries in our skies, he had the Enuma Elish rewritten. He interposed **his name** in place of our own sun's twin star (Nibiru). I shall not explain further. The proof is in my popular series **Exposed History of Planet Earth**. For those of you who have not read this series please accept the following as an explanation for the origin of Earth.

Science Facts

For many years it was widely accepted that our solar system contained nine planets. When Pluto exhibited erratic behaviour cosmologists preferred to downgrade it rather than have 'future egg on their faces'. Pluto is in fact a satellite of *another* star that is the twin of our sun.

This star has been called Planet X, Wormwood and Nibiru. Once again if you are unfamiliar with these names please read **Arrival of the Anunnaki.**

Suffice it to say astrophysicists have proven there is a second star that is our sun's twin. Of course the two stars cannot be close to each other. We would need Factor 1000000 suntan.

The two stars orbit in a figure of eight pattern (serpent-like as I shall explain later). Every 3600 years or so they reach their perigee (closest distance to each other).

Quote from space.com, August 26th 2013:

"Nemesis (Nibiru) is a theoretical dwarf star thought to be a companion to our sun. This theory was postulated to explain a perceived cycle of mass extinctions in Earth's history. Scientists speculated that such a star could affect the orbit of objects in the far outer solar system, sending them on a collision course with Earth".

Nibiru, Nemesis, Planet X, Wormwood is thought to be a brown dwarf. It is the sun's companion. There is strong evidence to show that when the two stars approach each other it may spell trouble for our planet. Anyone remember the Great Flood?

On those occasions when the twin stars do approach each other a cosmic cataclysm is extremely probable. The gravitational pull of two stars in our solar system is sufficient to affect the orbits of the planets. But we don't care about the other planets, only Earth.

Let us call a 3600 year period a SAR. The Sumerians did. They spent long periods observing the 'heavens' searching for the return of Nibiru. Sumerians were told that their gods had originated from Nibiru. Who told them? The gods themselves. Which brings me smartly back to Anu.

Anu was the boss. He ruled from Nibiru and visited Earth (Ki to Sumerians) every 3600 years or so. His eldest sons were given our planet as a present. İn return they gave Anu and his spouse Antu all the gold they could carry in a big space basket. No really that is why the 'gods' spent so much time on Ki. They were raping our planet.

Enki (real name Ea) was granted rulership of all that lay beneath the ground and the oceans. His younger brother Enlil ruled the air, airways (radio transmission) and the land. Despite Enki being older Enlil was heir to the throne of Nibiru.

Nibiru is commonly thought to be an inhabited planet that orbits our sun's twin star. Recently the star itself has been given the name Nibiru. For a full explanation I beseech you to read Arrival of the Anunnaki as many readers already have.

We shall return to the present after the remaining evidence from the Epic of Creation.

Tiamat was originally the third planet from our sun. The Earth did not exist. Marduk, Wormwood or Nibiru is reference to the sun's twin star and its orbiting planetary system:

As Tiamat opened her mouth to its full extent,
He (Marduk) drove in the evil wind. The terrible winds filled her belly, her courage was taken from her and her mouth she opened wide. He seized the spear and burst her belly. He severed her inward parts, he pierced her heart. He overcame her and cut off her life. He cast down her body and stood upon it.

When he had slain Tiamat her host was scattered…..
Kingu was exalted over all other warriors.

I have no intention of boring you with the enormously long Enuma Elish. Suffice it to say Tiamat was battered by the brown dwarf star system racing around our sun as it accelerated to break free of our sun's gravitational pull. In the process 'Nibiru' dragged its own planetary system in its wake as well as comets, electrical discharge, storms, etc

This cosmic cataclysm occurred around four billion years ago.

Tiamat was struck by seven heavenly bodies. It was ripped asunder (mouth she opened wide). Most of Tiamat (burst her belly) lies as the asteroid belt between Jupiter and Mars (the Firmament in Genesis). The much smaller chunk that remains to this day we call Earth.

One of the satellites that remained behind after this cosmic event was Kingu (this may indeed be our Moon).

Unbelievable? Then why are the American Military building a vast underground network that stretches from the west to the east coast of the USA?

Why are so many telescopes in space turned towards a particular celestial object that may be seen in the Southern Hemisphere right next to our sun? Nibiru is due to come closest to Earth in the year 2045. The elite want to be secure miles underground whilst the world's population fries.

We rely heavily on scientific proof these days. When scientific fact is not available there is a lazy tendency to return to religious teachings or Darwinism.

The major world religions are relatively young and each mimics the other in many ways. Despite the fact that Muslims, Jews and Christians are thought to have vastly differing belief systems it is merely the pomp and ceremony that separates them. The answer to how life was formed on Earth must be found in ancient knowledge (original religion) backed by modern science.

First the Modern Science explanation

"We want to know where we were born. If we can figure out in what part of the galaxy the sun formed, we can constrain conditions on the early solar system. That could help us understand why we are here," said Ivan Ramirez of The University of Texas.

2500 light years from Earth are three planets orbiting a brown dwarf. This star is almost identical to our Sun and is approximately the same age. Visit Daily Mail Science page.

One planet is a third of the mass of Jupiter and as it orbits its sun every 7 days it is so hot it appears to glow red with 'fiery wings'. This symbol is found in ancient archaelogical sites:

Nibiru is thought to return to our solar system every 3600 years. If this brown dwarf and its planetary system are indeed Nibiru then the time period is approximately correct.

Our sun and Nibiru orbit in a figure of eight around each other, much like a serpent.

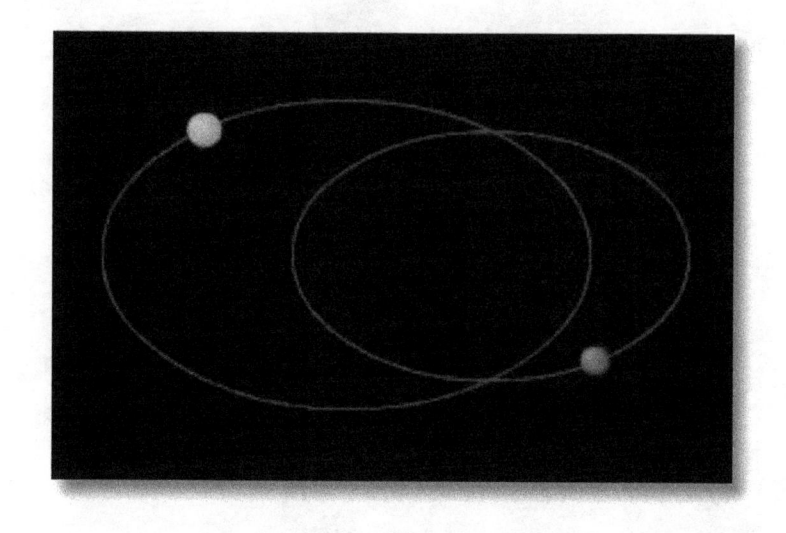

We shall investigate this comparison later.

Astronomers have used ESO's HARPS planet hunter in Chile, along with other telescopes around the world, to discover three planets orbiting a star in the cluster Messier 67. Pictured below is an artist's impression of a planet with a sun like our own.

This star is the first ever found that was created in the same "birth cluster" as our sun.

The chemical composition of this system is largely silicon, aluminium, magnesium and calcium. The elements and ratios are indeed similar to those found in our planet. Therefore, astronomers believe that one of the planets may be capable of supporting life.

Nibiru returns periodically dragging planets, satellites and space debris in its wake. It has brought devastation to our system on many occasions. One such occasion was recorded as the Great

Flood. Others may have signalled the demise of Lemuris and Atlantis. Read my book **Continent of Atlantis**.

The reason astronomers are so worried about the increased frequency of comets recently is because they know the sun's twin star has returned to our solar system.

As long ago as 1846 astronomers spotted Uranus behaving erratically. Uranus is unusual in that its spin axis is inclined by 98 degrees compared to its orbital plane around the sun. This is far more pronounced than other planets, such as Jupiter (3 degrees), Earth (23 degrees), or Saturn and Neptune (29 degrees). Uranus is, in effect, spinning on its side. Pluto has an erratic orbit that takes it around and between Neptune and Uranus.

July 2013, a report stated:
Neptune has a new moon, and its existence is an enigma. The object,

known for now as S/2004 N1, is the first Neptunian moon to be found in a decade. Its diminutive size raises questions as to how it survived the *chaos* thought to have created the giant planet's other moons.

Uranus's highly tilted axis makes it something of an oddball in our solar system. The accepted wisdom is that Uranus was knocked on its side by a single large impact, but new research revises our theories of how Uranus became so tilted and also solves fresh mysteries about the position and orbits of its moons.

By using simulations of planetary formation and collisions, it appears that early in its life Uranus experienced a *succession of small punches* instead of a single knock-out blow. This research has important ramifications on our theories of giant planet formation.

Most of the planets in our solar system spin on their respective axes "counterclockwise." But Venus and Uranus spin "clockwise" (if viewed from above Earth's North Pole, down onto the plane). Uranus is tipped at 90 degrees, spinning contrary to all of the other planets. Pluto is also tipped over, with an orbit that is mostly outside of the plane used by the other bodies orbiting the sun.

Do we not find these facts strange? Please investigate further.

My hypothesis is that our planet is the remnant of a far bigger planet (Tiamat) from a collision perhaps 14 billion years ago. The Pacific Ocean is testament to this as it appears as a yawning hole on one side of Earth where the remainder of Tiamat should be.

Microscopic lifeforms were most probably brought to Tiamat/ Earth during the collision with not one but seven heavenly bodies. Now we have a reason for why there are a myriad lifeforms on Earth. One planet is provided with a finite number of elements yet collectively a group of heavenly bodies would certainly contain a much greater *variety* of elements. Our planet was blessed with seven times the average number of elements granted to one planet.

This 'bath' of chemicals churning in super heated oceans goes a long way to support scientist's origin of life on Earth. In particular it answers the question "why is there so much *diverse* life on our planet?".

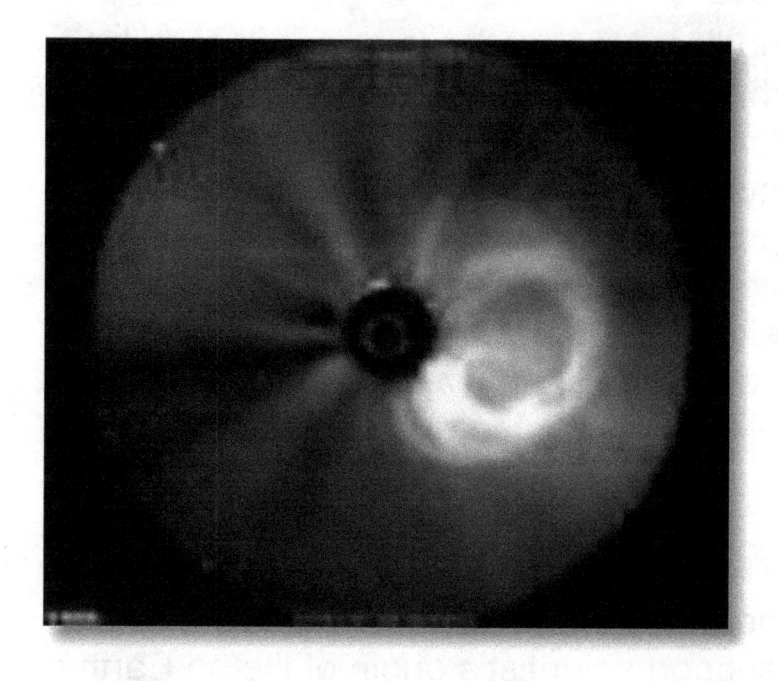

Recently astronomers have agreed that there must be another heavenly body way out there that cannot be seen. Not that it is too far away but that it does not reflect light easily. Hence they believe it to be a brown dwarf.

The solar system is now thought to have two stars: "the companion to our sun is a brown dwarf star or massive planet of mass between two and six times the mass of Jupiter".

A brown dwarf is a star too small to sustain the nuclear fusion that powers our sun and so is 'relatively' cool. How cool because I believe it may be millions of degrees Kelvin.

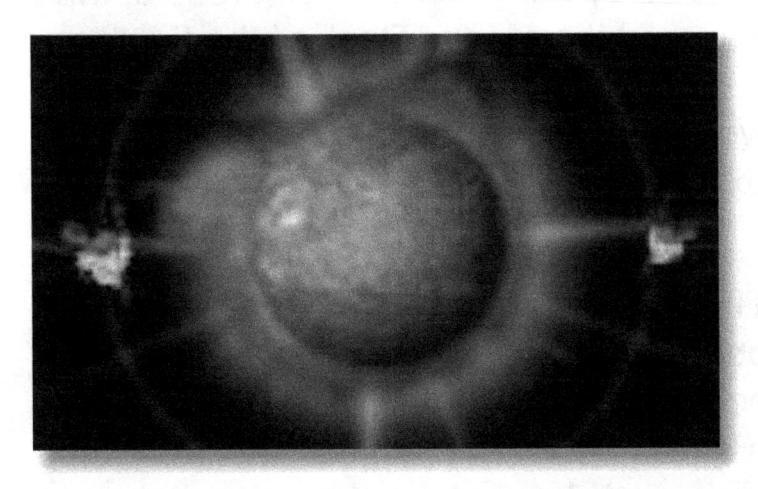

Ancient Astrology

Astrology has been derided in recent times in favour of astronomy. Astronomers are able to see the stars and planets and Galaxies and thanks to Sir Isaac Newton can plot the Precession of the Equinoxes.

Precession of the Equinoxes

The precession of the equinoxes is caused by the gravitational forces of the Sun and the Moon. Think of the Earth as a top spinning on its axis. Our planet wobbles just like the spinning top and the axis alters over time.

The Earth is not a perfect sphere but an oblate spheroid. If the Earth were a perfect sphere, there would be no precession. Equatorial diameter is 43 kilometers larger than the polar diameter. The applied forces of the sun and the Moon are almost perpendicular to the axis of rotation of the Earth.

The Earth is tilted on its axis. Therefore during much of the year the half of this bulge that is closest to the sun is *not* centred. The other half is off-centre on the opposite side. Since gravity decreases with distance the gravitational pull on the closer half is stronger. This imposes a torque on the Earth owing to the sun influencing one side of the Earth far more than the other. The axis of this torque is roughly perpendicular to the axis of the Earth's rotation. Therefore the axis of rotation precesses.

This average torque is perpendicular to the direction in which the rotation axis is tilted away from the ecliptic pole, so that it does not change the axial tilt itself. The magnitude of the torque from the sun (or the moon) varies with the gravitational object's alignment with the Earth's spin axis and approaches zero when it is orthogonal.

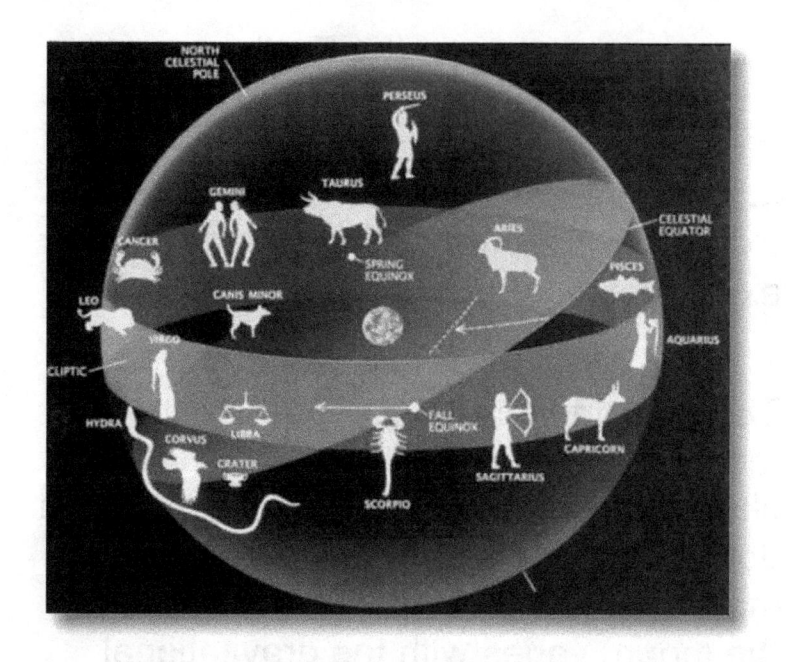

The effect of the sun and the moon on our planet is called the luni-solar precession.

Cosmologists have proven that not only do planets orbit the sun but our entire solar system gyrates in a spherical manner as it is influenced by gravitational forces from larger heavenly bodies. The sun and its respective planets orbit another star system which in turn orbits another and so on. In this manner our tiny

Earth is dragged around the outskirts of the Milky Way. It is now known that we complete a full orbit of the Milky Way every 25,700 years or so.

The Age of Aquarius is especially important as our planet and all the 'energy bodies' that exist on Earth are receiving a 'supercharge' of Universal Energy. In short we are leaving the 'Cosmic Winter' behind and entering the 'Cosmic Spring'. The Maya knew this as did many of our ancestors throughout the world.

The Mayan Prediction (incorrectly interpreted) states that in 2012 the planet enters a New Age. Beware listening to New Age Prophecy as much is self-styled aggrandisement. There are perhaps beings from our Galaxy or our future that have arrived in our skies at this auspicious time. There certainly is testimony from many people to contact with E.B.E.s – many of whom are very similar in appearance to humans. I am asking that you be careful before you believe the many 'trance state' messages that are 'being received' by self-styled mediums.

I shall now offer (in brief) testimony from a number of perhaps reliable sources. All insist that Nordics or Pleiadians are present and wish to aid humans in their transition to a higher dimensional existence. What if this was true? Wouldn't it be awesome.

Billy Meier Befriends Aliens

Born Eduard Meier on February 3rd 1937, living in Switzerland, this unassuming personality claims to have spent his life in contact with beings from the Pleiades. The Pleiadian system is commonly called the Seven Sisters Star system.

In Billy's possession are an incredible stockpile of photographs of UFOs. Some are flying others are on the ground.

Billy has travelled to many countries under instruction from first Semjase then the Pleiadian known as Quetzal. The latter is the youngest son of Enki – the Serpent who taught humanity.

In a television interview conducted by Nippon Television (Japan) in 1980, Meier is asked about the attempts on his life. On the evening of May 10, 1980 at 10 p.m a shot was fired at Billy. The bullet missed Meier's head by only 203 mm. The bullet was found embedded in a concrete wall directly behind the couch on which Billy was sitting. This was not the first attempt on his life. One occurred in Hinwil during the Winter some time between 7 and 8 p.m. A bullet pierced Billy's office window, closely missing his head. The thrid attempt was partially successful in that the bullet entered his chest. For years Billy has existed with only one arm after yet another 'accident'.

If the accounts of Billy Meier are fabricated who and why should want him dead?

Some of the information presented by Pleiadians over the many years of contact:

- Our sun has a "dark twin" which is located about one light year away. Therefore we live in a binary Star system
- There are 280 chemical elements in the universe, but it is a creational law that no

single Star contains all 280 of them.

- The next inhabited solar system is around five light years away from Earth. Different worlds in that system are inhabited by human forms of life. One of the planets located there is called Akart.
- There was another planet in our solar system called Malona, which was destroyed by its inhabitants. The asteroid belt is the remnants of that former planet.
- Our universe is around 46 trillion years old, not billions of years old, as Earth scientists claim.

Meier's alleged discussions with the Plejarens are highly detailed and wide-ranging, dealing with subjects ranging from spirituality and the afterlife to the dangers of mainstream religions, human history, science and astronomical phenomena, ecology and environmental dangers, in addition to prophecies of future historic trends and events.

An additional aspect of the Meier case is the highly controversial book the Talmud Jmmanuel. It is said to be the translation of ancient Aramaic scrolls that were discovered by Meier and a colleague in Jerusalem in 1963. The book claims to be the original teachings and life events of the man named Jmmanuel (Jesus Christ). Meier claims to have travelled back in time to meet Jmmanuel and his wife Maryam.

I have written several books detailing the wife and family of Jesus. Feel free to read them.

There is a plethora of information and testimony regarding Meier's meetings with extra terrestrials. I suggest you check it out for yourself and reach your own conclusion.

Dawning of a New Age

We are entering the Age of Aquarius as we leave the Age of Pisces. The Age of Pisces coincidentally began around the time of the birth of Esau (Jesus Christ). According to astrological mysticism, there will be unusual harmony and understanding in the world. Those who follow that belief system see it as a turning point in human consciousness in which balance is restored by consciously moving beyond the physical body.

Nasa is plotting the return of our twin 'Sun' and its orbiting planets. Observatories around and above our planet are scanning the heavens for extremely important data regarding the Red Planet that creates cosmic havoc every 3600 years as it plunges through our Solar System.

The USA has bled the country dry to fashion a vast underground complex that stretches from the West to the East coast. When the expected cataclysm occurs the elite will live in this incredible complex whilst the remaining billions perish.

Of course this is the reason for all of the ancient 'observatories' from Stonehenge to Macchu Picchu. They were created by our ancient 'ancestors' to plot the

Pregression of the Equinoxes and predict the return of our binary Star system.

Natural disasters occur in cycles and these cycles correspond to the reappearance of Nibiru as detailed in the ancient Sumerian Scriptures. I have written extensively on this subject.

Why are you a Christian, Jew, Hindu, Sikh, Atheist or Muslim. Most people state that it is their belief yet in over 90% of cases it is because your parents reared you to believe what they themselves were reared to believe.

Whereas none of the belief systems are entirely wrong there are fundamental flaws in all belief systems especially if the believer has accepted their doctrine at face value.

The Truth lies in all religions and also in the belief that we are all aspects of the One. In fact we believe we are the physical body when the truth is we inhabit the body for a lifetime. Think of the physical body as an overcoat. When it wears out we simply discard it in favour of something better. This process is known as death. The majority of humans avoid the subject of death until they finally must face the reality.

Physicists have informed us that our bodies are merely clumps of atoms. These atoms are so far away from each other that if one was the size of an orange its closest neighbour would be a football field away. Pursuing this analogy the space inside an atom is so vast that the entire Universe would fit neatly inside one atom should it be possible to shrink the Universe to an infinitely dense particle. Superstring theorists have envisaged over eleven dimensions all co-existing with an infinite number of Universes. This is explained in full in my extremely popular series **Worship of the**

Because we can perceive light in the visible spectrum and our hearing range is equally impaired humans are unable to 'see' beings that are present in the same space and time as we exist. Animals on the other hand do see such beings and are affected by their presence.

Laura Eisenhower is the great granddaughter of Dwight D Eisenhower. Andrew Basiago and Laura were placed in a black ops program when they were young children.

Andrew Basiago claims to have been entered into the DARPA time travel Project Pegasus between the years of 1968 and 1972. Initially the 'trips' were between two CIA facilities, based in New Jersey and El Segundo California.

Many years earlier the genius Nikola Tesla had been murdered by CIA operatives. His many inventions have been used by the US Shadow Government ever since. HAARP originated with Tesla (although he wished to use his idea to provide the world with free energy).
One of his other amazing inventions was a means to travel instantly through space or time.

After Tesla's murder in January 1943 his documents and equipment were seized by the CIA. One document was a working schematic for a teleportation machine. Basiago states that "radiant energy" is utilised to power the time/teleportation apparatus. The machine forms a "shimmering curtain" between two elliptical booms.

Radiant energy is pervasive throughout the universe. It is unlimited and therefore free energy. One of its abilities is to bend time-space. The general public are unaware of the possible uses of this free energy as the oil barons and US Military have no intention of providing the world with it.

Basiago claimed to have passed through a curtain of energy into a vortal tunnel. The process would seem tol ast around forty five seconds whereupon he would find himself at his destination. He made hundreds of trips some off planet.

Other experimental teleportation devices included a plasma confinement chamber in New Jersey and a room in El Segundo, California used for the jumps. He describes a form of holographic technology that allowed him and other children to travel physically and virtually.

Below is a photograph taken in Gettysburg in 1863 when Lincoln addressed the nation.

Basiago claims to be the boy featured prominently in the foreground.

"I had been dressed in period clothing, as a Union bugle boy," he said. "I attracted so much attention at the Lincoln speech site at Gettysburg -- wearing over-sized men's street shoes -- that I left the area around the dais and walked about 100 paces over to where I was photographed in the Josephine Cogg image of Lincoln at Gettysburg."

Basiago claims on several occasions he was sent back in time to Ford's Theatre the night of Lincoln's assassination.

"I did not, however, witness the assassination," he said. "Once, I was on the theater level when he was shot and I heard the shot followed by a great commotion that arose from the crowd. It was terrible to hear."

"As these visits began to accumulate I twice ran into myself during two different visits."

These are astounding claims yet Basiago goes on to say that there is a human colony on Mars.

Andrew and Laura both agree that the Black Ops Budget is in part subsidising travel to and from Mars and the stockpiling of provisions and creation of cities on that planet. Basiago claims he was sent to Mars on several occasions.

Michael Relfe: Mars colony eyewitness Michael Relfe is a whistleblower and a former member of the U.S. armed forces who in 1976 was recruited as a permanent member of the secret Mars colony. In 1976 (Earth time), he teleported to the Mars colony and spent 20 years as a permanent member of its staff. In 1996 (Mars time), Mr. Relfe was time-traveled via teleportation and age-regressed 20 years, landing back at a U.S. military base in 1976 (Earth time). Mr. Relfe then served six years in the U.S. military on Earth before being honorably discharged in 1982.

One of the other time travellers in Basiago's class was Barry Soetoro. We know Barry today as Barack Obama.

Alfred Webre, a lawyer specializing in exopolitics (political implications surrounding an extraterrestrial presence on Earth) substantiates Basiago's claims. Webre insists teleportation and time travel have been practiced on human 'volunteers' for at least 40 years.

"It's an inexpensive, environmentally friendly means of transportation. The Defense Department has had it for 40 years and Donald Rumsfeld used it to transport troops to battle."

It is a positive thing for the human race to rear colonies on other planets. Earth has suffered multiple cataclysms in the past.

We should protect the human genome by placing human settlements on other celestial bodies.

"Yet, when secrecy surrounding such projects tempts government to rob the free will of individuals, and excludes humanity from debating a subject that implicates the whole human future, and diverts the destiny of a planet to serve an off-planet agenda, the conscience of a free people requires that such projects be undertaken in the bright sunshine of public scrutiny, not within the dark corridors of the military-industrial complex".   – Andrew D. Basiago and Laura Magdalene Eisenhower, 2010

Teleportation and Levitation

It is now widely accepted amongst mainstream science and archaeology that the Egyptians could not possibly have erected the pyramids using slave labour alone. In fact as we Google Earth our way around the planet pyramids and ziggurats pop up on almost every Continent.

So how were all the major ancient sites constructed (please research the fantastic statues and cities almost forgotten in South East Asia – places such as Cambodia)?

The current theory involves both levitation and teleportation.

Levitation is the raising of an object (often through the power of the mind) which ordinarily would not be possible by such means (eg a pen with sheer will) or is too darned heavy to ascend with even the most powerful pieces of kit. Take for example the site at Baalbek in Lebanon:

This site was first mentioned in the Epic of Gilgamesh

The smallest foundation Stones at Baalbek weigh over 100 tons each.

No machine on Earth is able to lift or carry even one of these stones let alone transport the entire (some say space craft landing site) a distance of 25 miles from the quarry from which they occurred naturally. So perhaps levitation or even teleportation is the only possible solution. What is teleportation? Perhaps you have seen old episodes of Star Trek?

Teleportation is a 'hypothetical' mode of instantaneous transportation; matter is dematerialized at one place and recreated at another. Andrew Basiago will tell you it is not hypothetical.

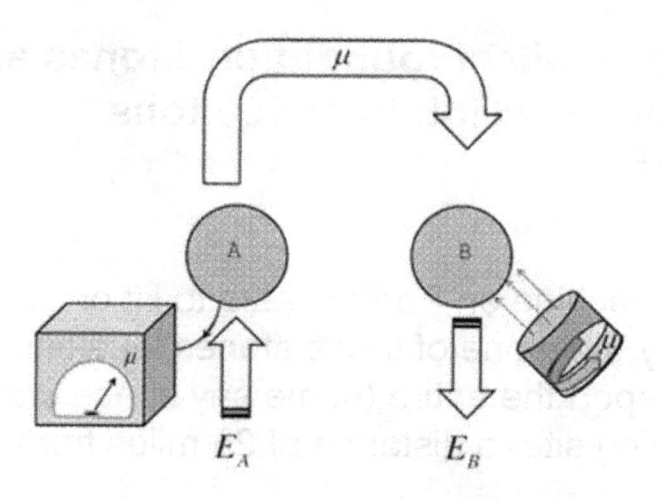

Teleportation's As Easy As 1-2-3!

Over five years ago, scientists succeeded in teleporting information. Unfortunately, the advance failed to bring us any closer to the *Star Trek* future we all dream of. Now, researchers in Japan have used the same principles to prove that energy can be teleported in the same fashion as information. Rather than just hastening the dawn of quantum computing, this development could lead to practical, significant changes in energy distribution.

According to the theory, developed by Masahiro Hotta of Tohoku University, Japan, a series of entangled particles could be stretched across an infinite amount of space. By inducing an energy change in one of the particles, the other entangled particles would change as well. Eventually, to preserve conservation of energy, the original particle would be destroyed, with its energy passing to the final particle in the chain. Thus, the energy has been teleported from one particle to another. Just like Star Trek. Our bodies are after all a form of energy.

The following is paraphrased from jcvi.org:

In 1995, a group of scientists from the J. Craig Venter Institute, JCVI, launched a project "*minimal genome*" whose goal was to leave in the genetic code only the essential information for the existence of a living organism. The research was based on the bacterium *Mycoplasma genitalium,* which at that time had the smallest genome of all known (482 genes). In 1999, Venter reduced the number of genes to 382, creating a "semi-synthetic" Mycoplasma laboratorium.

In 2003, together with colleagues from JCVI he announced to have managed to transform one type of bacteria (*M. capricolum*) to another (*M. mycoides*) using genome transplant, and in 2008 an organism with a fully synthetic genome was created after a second attempt.

Last year, Craig Venter published a new book Life at the Speed of Light: From the Double Helix to the Dawn of Digital Life, in which not only he describes the history of synthetic biology but also a completely new concept of "*biological teleportation*".

Venter argues that the information contained in the DNA can be digitized and transmitted over cable or via electromagnetic waves, and on the other end a special machine would synthesize oligonucleotides (short DNA fragments) and a robot would assemble the DNA completely identical to the original.

Such a "teleportation" promises significant savings of resources: in fact, to transmit information about the status of all the atoms in the human body about 1032 bits are required, but for the transmission of digitized human genome – only 6 x 109 bits.

Technology that Venter writes about is not fiction – the company *Synthetic Genomics,* founded by Venter and his colleague Hamilton Smith, has been working on reading and assembling the DNA. But, of course, teleportation of a person in this manner cannot take place either now or in the near future. But to transmit information about the strain of a new virus from one laboratory to another and quickly create a vaccine to prevent pandemic can be possible now.

In 2011, *Synthetic Genomics* in collaboration with the pharmaceutical company *Novartis* demonstrated an "assembly" of a vaccine based on the envelope proteins of avian influenza virus H7N9 (later the method was repeated for other strains) in less than five days (usually a similar vaccine production requires about two months).

Moreover, such method can allow to build specialized bacteriophages which can be an excellent alternative to the use of antibiotics.

However, Venter claims that these methods can be used not only on our planet. Why not to use "biological teleportation" to detect life on Mars? Equipping rover with a biological laboratory, it is possible to transfer information about the found genomes to Earth, where they can be recreated and studied. And maybe someday the process will be reversed: the colonists will be able to receive the necessary bacteria and plants from Earth.

Who thought that the 3-D printer would become a reality?

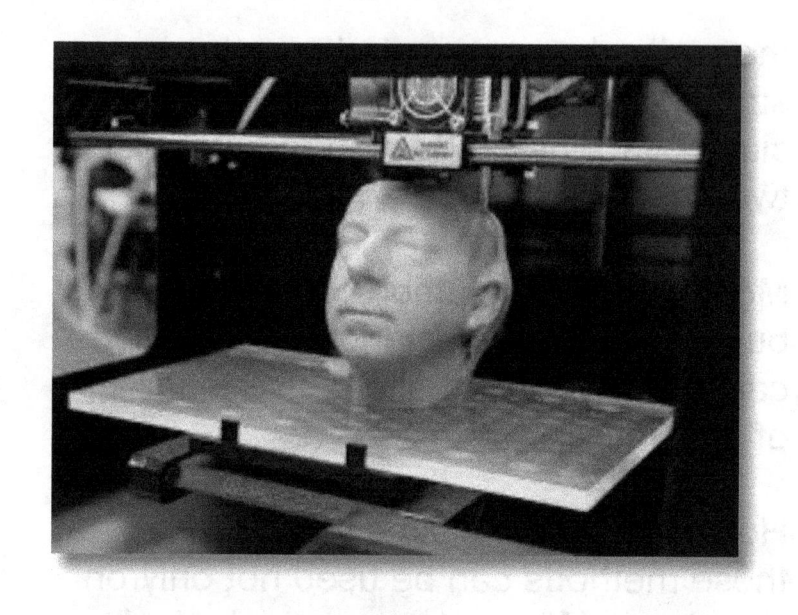

Quantum Teleportation

What is teleportation? Here is an example. Lab A and Lab B each have a box. Teleportation should transport any object that is placed in Box A and move it to Box B.

Of special interest to science fiction fans (among others) is human teleportation, where a brave telenaut (whom we shall call Jim) enters Box A and uses the teleportation machine to travel to Lab B. It turns out that human teleportation appears possible in principle.

Teleportation of much smaller objects is not only possible but has been accomplished in the laboratory. Our goal here is to explain how teleportation is achieved.

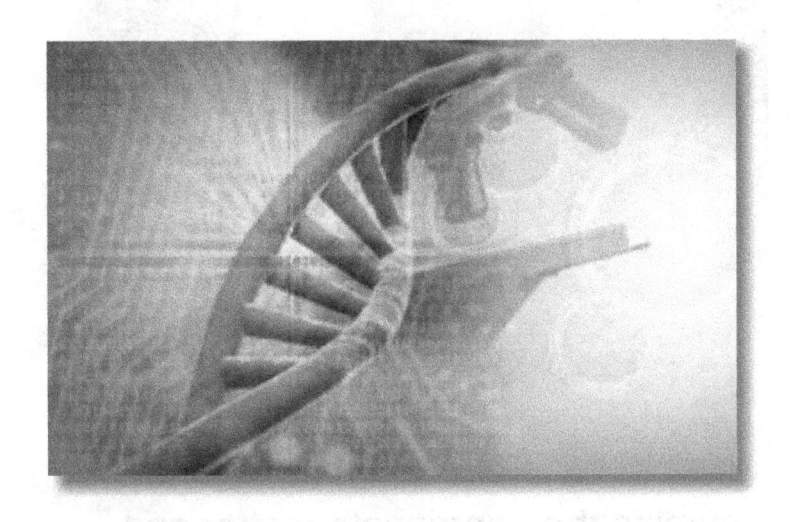

Classical teleportation

Quote from link.aps.org:

Load the cargo of box A onto a truck, drive the truck over to lab B then unload the cargo into box B. Of course this is transportation not teleportation.

A giant 3-D fax machine could in theory reproduce a human being.

Assuming the appropriate matter and energy is available in Lab B. The aim is to assemble it such that it is identical to the object placed in Box A.

There are so many drawbacks to this method not least that there would be two identical humans then three and so on.

Quantum teleportation

The surprising result of quantum teleportation is that even though the "measure and reconstruct" procedure does not work, there is an alternative procedure that effectively validates teleportation in the quantum world.

Not until the publication of a 1993 paper by Bennett, Brassard, Crepeau, Jozsa, Peres and Wootters that scientists realized quantum teleportation was possible.

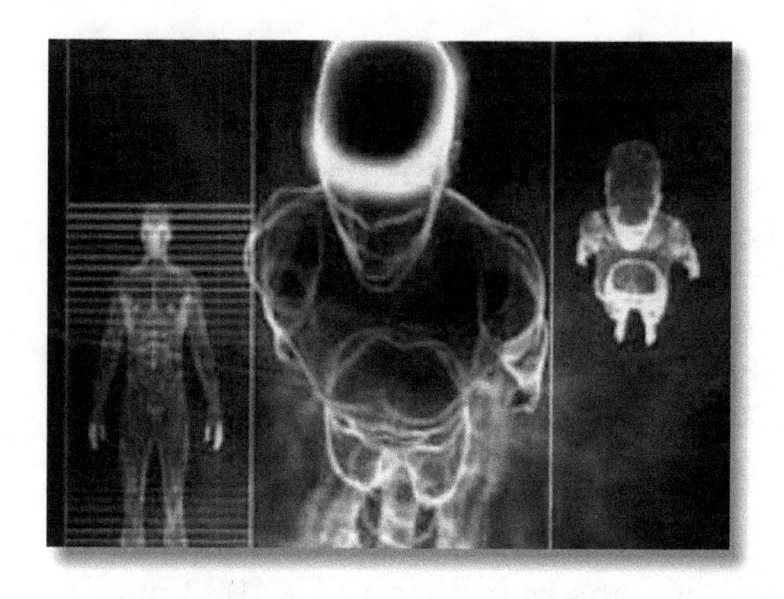

An entangled state is a pair of objects that are correlated in a quantum way. Below we will describe a specific example known as the "singlet state" of two spins. However, let us first explore the consequences of this extra requirement for quantum teleportation.

To prepare an entangled state of two particles, one essentially has to begin with both particles in the same laboratory. Begin in Lab A.

In practice sending this particle to Lab B will destroy one entangled state to create another entangled state.

The only solution is that in the past the wall that separates Lab A and Lab B was non existent. At that time scientists from the two labs colluded, created a large number of entangled states, then carried them to their respective laboratories.

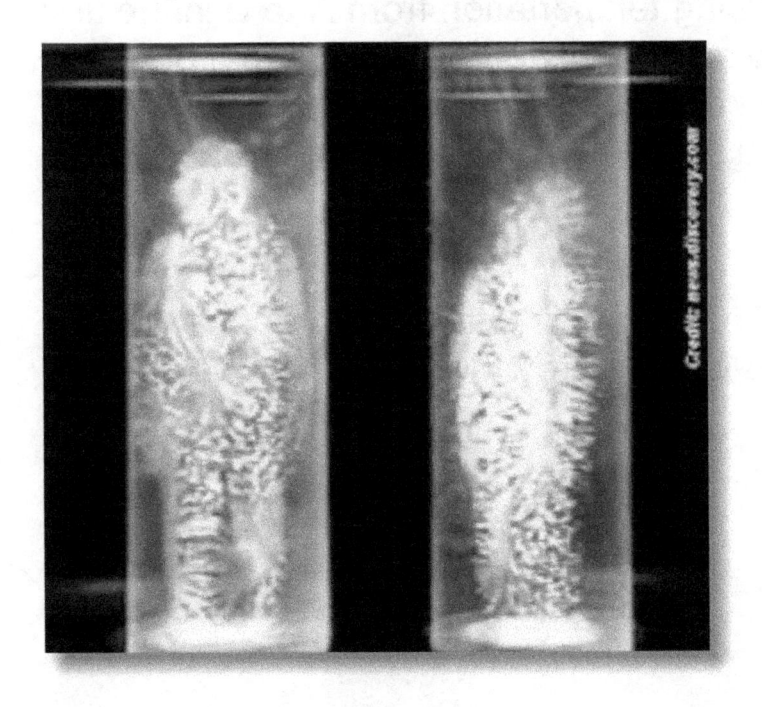

The entanglement requirement poses a second problem since it is destroyed when used.

Entanglement is effectively a resource that is slowly depleted as teleportations occur. It can be renewed by meeting in person and then carrying entanglement back from Lab A to Lab B but it has to be transported without the use of teleportation. In principle this is difficult, otherwise we wouldn't have bothered using teleportation from A to B in the first place. However, the idea is that one difficult journey from A to B will allow in future many quick transfers from A to B.

Quantum information cannot be copied. The only way to teleport an object to Lab B is to destroy the object at Lab A.

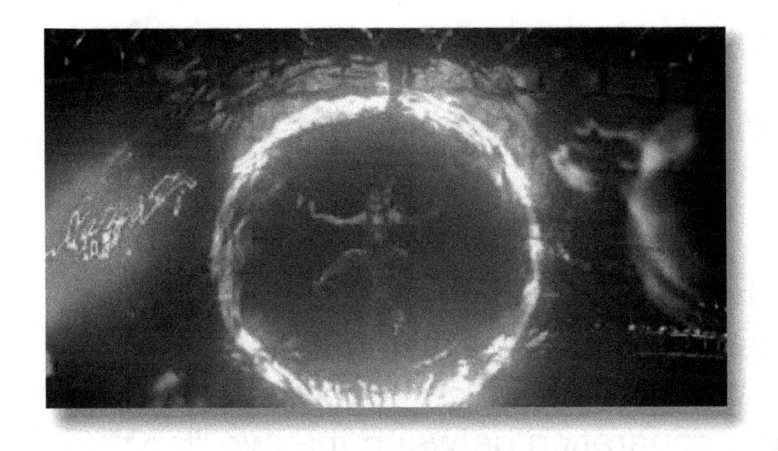

At first glance, though, there seems to be a way to use the teleportation procedure for superluminal communication. That is, by measuring the spins in Lab A, we are instantaneously modifying the spin in Lab B.

Real experiments that perform teleportation

A number of groups conducted experimental realizations of the quantum teleportation procedure described above in the years 1997 and 1998, using a variety of different systems such as the spin (or polarization) of photons and the spin of atoms. In many cases Labs A and B were the left and right side of a table, and the spins were teleported roughly 50 cm.

The reason distance becomes relevant has to do with the distribution of entanglement which becomes harder as the separation between the two "labs" increases.

A second related problem is the storing of entanglement which can only be done for very short periods, so in practice most early experiments distribute the entanglement only moments before it is to be used for teleportation. However, these experiments were sufficient to convince most physicists that teleportation of spins is possible.

Since 1997 there have also been many improved versions of the teleportation experiment. For instance, the distance has been increased in one experiment to

600 m, and the accuracy of the teleported state has also been slowly improving.

In principle, if you can teleport one spin, then you can teleport many spins simply by repeating the experiment in series many times. But this roughly only works on disjoint spins.

To teleport a single object comprised of many spins is still out of reach of present day experiments.

In the future, though, we should see experiments that teleport large numbers of spins. Certainly, if a practical quantum computer is ever built then the same technology would likely allow us to teleport a few thousand spins. It is likely that this will happen within the next 30--50 years, if not sooner.

But will we ever be able to teleport people?

There are some 10^29 matter particles comprising a human person, each of which has position and momentum degrees of freedom in addition to spin. In principle, we might also need to teleport the photons, gluons and other energy particles comprising a person.

While most scientists expect that ten, hundreds and maybe even thousands of spins will be teleported in practice some day, the teleportation of a human being, even in principle, is actually still a controversial subject.

In the meantime, one can ask if there any applications for teleporting thousands of spins?

In the future it is likely that quantum computers (i.e., computers capable of processing quantum information) will be built. These computers will need to exchange quantum information. One way these exchanges of information can occur is via a quantum phone (a device capable of sending and received quantum messages). Some time within the next 100 years you will see a quantum teleportation device for sale in your local computer store.

Holograms

The dictionary definition of virtual reality: The computer-generated simulation of a 3-D image or environment that can be interacted with in a seemingly real or physical way by a person using special electronic equipment (such as a helmet with a screen inside or gloves fitted with sensors).

This interaction with the environment is unreal in the respect that the objects and encounters are not present in our world yet fool the senses into believing they are real. This brings me a very important point. What if the entire physical world around us is simply a super cosmic computer generated image? Next time you are immersed in a real time computer game with others pause for a moment and think how real this all seems at that moment.

Virtual Reality as a means to escape for purposes of cerebral adventure is still in its infancy. Problems with the cumbersome headgear and nausea need to be addressed. However, in time, we will all be acting out our fantasies with over the counter products. For now I wish to point you in the direction of existing uses of VR that are invaluable to science.

VR is used extensively in the military to train troops and also many suspect a more sinister purpose. Gamers are enlisted to 'take out' targets they initially believe to be computer generated personnel. It transpires that the targets are live defenceless civilians from foreign soils.

VR in healthcare is a real boon. Virtual autopsies are performed as are virtual operations. In this way student doctors (especially surgeons) may practice on virtual patients until they become proficient.

Rolls Royce engines and turbines and many other similar products are now designed using Virtual Reality techniques.

VR is entering the field of education as more and more academics realise the value of 3-D.

Scientific projects are often trialed using VR.

Other uses of VR are found in movies, construction, sport, programming languages and telecommunications.

I refer you to http://www.vrs.org.uk/virtual-reality-applications/index.html and similar websites.

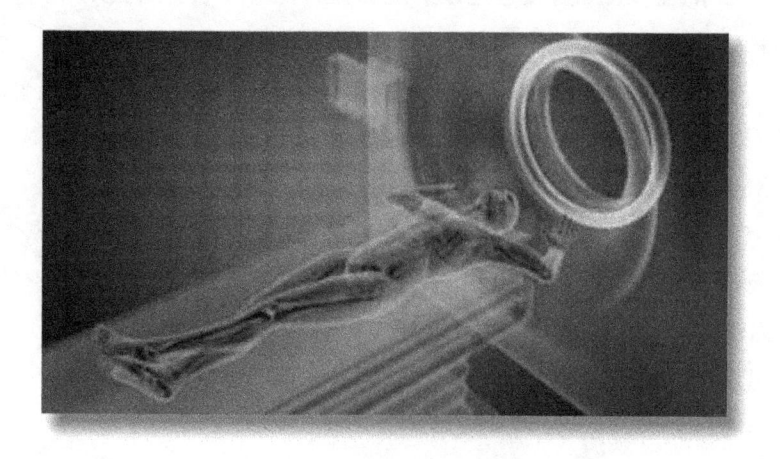

Now that we possess knowledge that our Universe and our own bodies may indeed be a massive hologram let us briefly look at the explanation: superstrings.

Cosmic Strings

Cosmic strings are thin threads of energy that are hugely dense. The tiniest string is able to warp time. If two infinite strings passed each other they effectively create a time machine by warping light eccentrically.

Cosmic Strings are hypothetically a 1-dimensional (spatially) topological defect in the fabric of spacetime left over from the formation of the universe. Interaction will create fields of closed time-like curves permitting backwards time travel.

Some scientists have suggested using "cosmic strings" to construct a time machine. By manoeuvering two cosmic strings close together – or possibly just one string plus a black hole – it is theoretically possible to create a whole array of "closed time-like curves." Cause two infinitely long cosmic strings to pass each other at very high speeds. Now if a spaceship should fly in a figure of eight around them it will be able to travel *anywhere* in the Universe.

A black hole contains a one-dimensional singularity (infinitely small point in the space-time continuum).

A cosmic string would be a two-dimensional infinitely thin line that has even stranger effects on the fabric of space and time. Although no one has discovered a cosmic string astrophysicists have suggested that these strings may explain strange effects seen in distant galaxies.

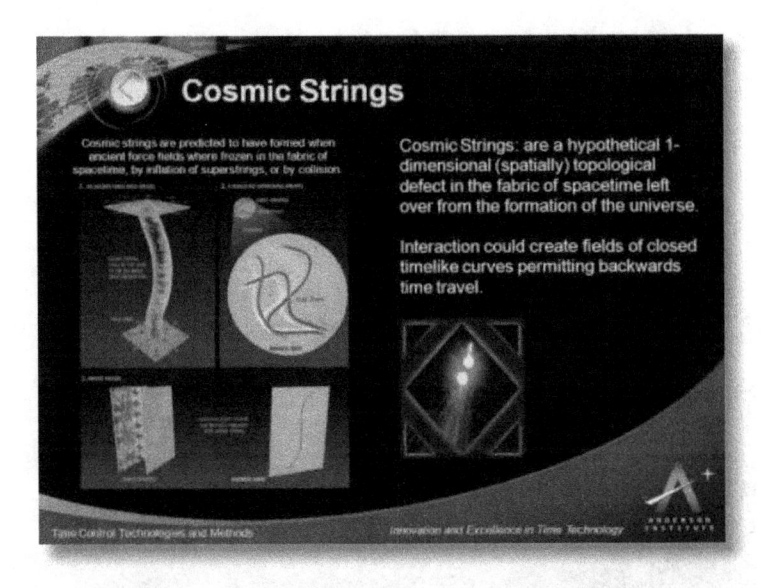

Cosmic strings are hypothesized to form when the field undergoes a phase change in different regions of spacetime, resulting in condensations of energy density at the boundaries between regions. This is somewhat analogous to the imperfections that form between crystal grains in solidifying liquids or the cracks that form when water freezes into ice.

The phase changes that produce cosmic strings may have occurred in the earliest moments of the universe's evolution.

The key characteristics of the application of cosmic strings for time control and time travel are presented in the picture below.

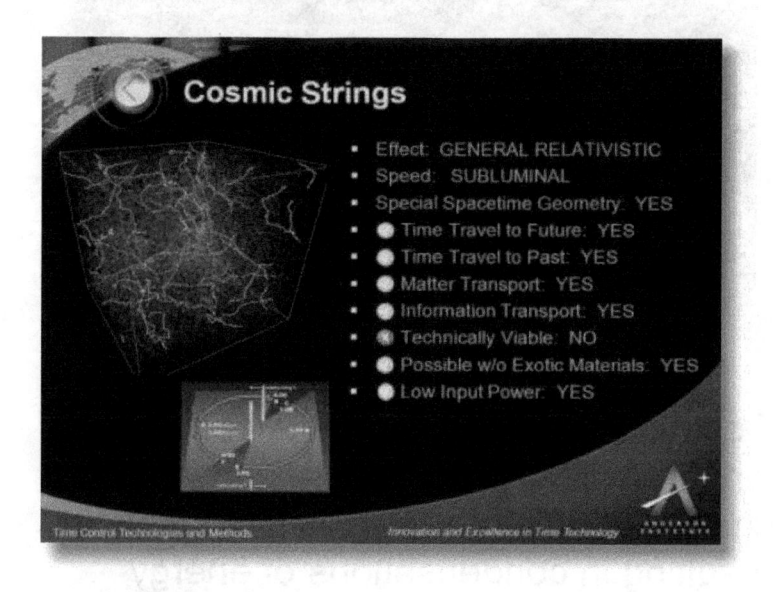

Cosmic Strings

- Effect: GENERAL RELATIVISTIC
- Speed: SUBLUMINAL
- Special Spacetime Geometry: YES
- ● Time Travel to Future: YES
- ● Time Travel to Past: YES
- ● Matter Transport: YES
- ● Information Transport: YES
- ● Technically Viable: NO
- ● Possible w/o Exotic Materials: YES
- ● Low Input Power: YES

Time Control Technologies and Methods — Innovation and Excellence in Time Technology — ANDERSON INSTITUTE

A cosmic string 1.6 kilometers in length may be heavier than the Earth. However general relativity predicts that the gravitational potential of a straight string vanishes. There is no gravitational force on static surrounding matter.

The only gravitational effect of a straight cosmic string is a relative deflection of matter (or light) passing the string on opposite sides (a purely topological effect). A closed loop of cosmic string gravitates in a more conventional way. During the expansion of the universe, cosmic strings would form a network of loops, and their gravity could have been responsible for the original clumping of matter into galactic superclusters.

A cosmic string oscillates at near the speed of light. This causes part of the string to pinch off into an isolated loop. Such loops have a finite lifespan due to decay via gravitational radiation.

On Earth a tunnel will be the shortest distance either through a mountain or through the Earth itself. In spacetime the shortest distance is through a wormhole:

Think of a guitar string that has been tuned by stretching the string under tension across the guitar. Different musical notes are produced by plucking the guitar string more vigorously or for a longer period.

Elementary particles (observed in particle accelerators such as the Large Hadron Collider) in string theory are akin to musical notes. These 'musical notes' are termed **excitation modes.**

Strings in string theory are floating in spacetime. To become excited the string must be stretched and sure enough each invisible string has tension. This tension (alpha prime) is precisely equivalent to the square of the string length scale.

It would be ideal if the average size of a string is approximately the same as the length scale of quantum gravity. This is the **Planck length (**about 10^{-33} centimetres, named after Max Planck)

String theories presuppose the strings to be closed loops or the particle spectrum includes fermions. To include fermions in string theory we must have a specific form of symmetry we call **supersymmetry.**

Supersymmetry relates all particles that transmit forces to particles that comprise matter.

For every boson (particle that transmits a force) there is a corresponding fermion (particle that comprises matter). The LHC and other particle accelerators are searching for evidence for high energy supersymmetry.

This whole concept is a means by which physicists are attempting to answer the many questions they have defining the

Universe in terms of Newtonian physics. Whereas once the Earth was flat and everything revolved around our planet then Newton introduced gravity to a spherical world that occupied a minute amount of space in an insignificant Solar System we realise we still do not have our facts right.

I refer you to http://www.superstringtheory.com/ and similar websites for a full explanation.

Serpent Mounds

Quote from Indianapolis Journal 5th March 1894:

Giants of Other Days: Recent Discoveries near Serpent Mound, Ohio.

Warren Cowen recently discovered several ancient graves in Highland County, Ohio. The Serpent Mound is one mile away. Limestone slabs covered each grave. Inside each grave was an extremely old skeleton. The remains were of 'humans' almost three metres in height.

This high point of land is known to locals as the Serpent Mound. The countryside around represents a serpent coiling several times with a distinct head and an 'egg'.

Situated in Ohio Brush Creek, Adams County, Ohio the Serpent is over 420 metres long, raised one metre from the ground. The head is aligned to the summer solstice. The curves of the body correspond to lunar movements. The coils are aligned to the solstices and the equinoxes. If 1070 CE is accurate as the construction year, building the mound could theoretically have been influenced by two astronomical events: the light from the supernova that created the Crab

Nebula in 1054, and the return of Halley's Comet in 1066

The Serpent Mound is thought to have been created in accordance with the pattern of stars known as the Draco Constellation.

The 'egg' is thought to be our Sun. The Serpent is depicted as 'swallowing' the egg.

This is some feat for 'primitive' humans.

More recently a crop circle has appeared that points to the Serpent Mound.

Who created it, when and why?

Yet there are many more such mounds throughout the world, all representative of a Serpent. These include the Avebury and Cadbury mounds in Wiltshire and the ancient site of Camelot in England.

When Stonehenge was first built it was most likely in the form of a Serpent. The famous standing stones are the head. Over the centuries many stones have been taken to build dwellings. Other stones are dotted around the countryside and only an aerial view and a little imagination shall confirm it is in the shape of a serpent.

All Serpent sites are on ley lines (power routes).

The Serpent is representative of the cosmological path (figure of eight) of the twin stars. The Cosmic Egg is our Sun being 'swallowed' by its twin every 3600 years. If you are in the Southern Hemisphere look at the early morning sky. There are two suns.

The Sacred Serpent

When St. Patrick sailed to Ireland he was on a rather strange mission from the Pope. St. Patrick had been instructed to rid the land of snakes. This is a task that no singular individual can accomplish and furthermore to my knowledge there were no poisonous snakes in Ireland ever. Was the Pope planning a summer holiday in Galway?

Like all ancient tales the truth masquerades as fiction. Snakes were not mentioned in the communication between the Pope and Patrick, rather it was serpents.

I believe I have mentioned the two sons of Anu previously. They were named Enki and Enlil. They and their families ruled the entire planet for hundreds of thousands of years. Their eldest sons were rivals for the throne of Nibiru. Marduk and Ninurta respectively battled all over our planet many many times. Marduk has many names including Zeus and Jupiter. Ninurta likewise was Mars and Aries. Again I point you to my previous writings.

The family of Enki were the 'Serpent Line' whereas Enlil's kin were bulls. Marduk chose the Ram as his symbol as he eventually singlehandedly ruled our planet (some believe he still does) when Aries was ascendent in the 'heavens'.

Symbology since ancient times abounds. I shall provide you with a few scant examples.

Quetzalcoatl (alias Ningizzida, Enki's youngest son)

The Feathered Serpent

Ram (eses), Ra, Zeus, Marduk, Amon are one and the same.

Ram symbol for Ra

Enlil (bull) ruled during the period when Taurus was in the ascendent.

Now we know why Astrology was so important to the ancients. The ascendency of each constellation signified the period of rulership for each chosen 'god'. Note the Zodiac goes 'backwards'.

Every single ancient civilisation relied on the Zodiac, standing stones, observatories and space satellites (yes the 'gods' built them long before we did) to ascertain just precisely where the Nibiru system was in relation to Earth.

The serpents that St. Patrick had to rid Ireland of were descendents of Enki. What? Gods in Ireland? Haven't the Irish people been telling you this for centuries. They were ruled by giants and these giants were gods. As an aside why not investigate Hy-Brazil (the remnants of Atlantis) also in the Irish Archipeligo. Not so stupid these Irish, eh? Why do you think the Black Prince and successive English Governments wanted to 'rape' Ireland?

Many authors (including myself) have proved conclusively that 'aliens' were our Overlords for hundreds of thousands of years. I am not going to repeat their work. These extra terrestrials originated from other planets.

What is Real Reality?

We blindly accept that our Governments are acting in our best interests. We follow our parents' religion like sheep. We identify ourselves by nationality, gender and sexual preference. We succumb t control by the authorities, police forces and armed forces. We cheer like football hooligans when our military might murders millions of innocent civilians. We feel safe behind the nuclear might of our superpower countries. Yet what if those same authorities, those same leaders, those same armies turned on us?

This is happening today in the USA. Given unlimited power the Shadow Government under instruction from a small group of the elite we know as the Illuminati are seizing our homes, our guns and our freedom. This is the Final Curtain for humanity unless some 'wildcard' is played by a force that we do not know exists.

The Matrix movies tell it like it is – examine us at molecular level we are empty spaces, holograms. The five senses operate to provide a 'sense' of reality.

We feel it therefore it must be real. We see it therefore it is visible. Yet outside the tiny spectrum that humans perceive there is much more ten dimensions more and more Universes and so on. In fact all that could possibly exist does exist simultaneously everywhere.

God is not coming to save us. We must save ourselves. Know that you will die (leave the body) and if you are not fully aware you shall be reborn (owing to desire for what is here in the illusion). By losing all desire for material things including relationships, wealth, body consciousness etc then and only then do you pass to the next dimension – step up in Reality.

Miracles happen all the time – silly ones where someone eats a light bulb or important ones where a mother lifts a car to save her trapped baby. People have walked through walls, been transported from one spot to another on a different continent and some have x-ray vision. Please read my books on this subject. Because these things are not mainstream we marvel at them then we 'go back to sleep'.

George Bush   TotallyLooksLike.com   Chimpanzee

But

What if the good and the evil is a juxtaposition of thoughts played out in our 'reality'?

What if all musical compositions were a Divine joint effort?

What if all inventions have been Divinely inspired?

What if we are simply drops in an Ocean?

What if we believed that we are all capable of miracles *all* of the time?

What if we really knew that we are ONE consciousness examining itself.

What if we are all One buying in to the Cosmic Computer Game?

The entire fabric of our 'reality' on planet Earth is like the pack of cards in Alice in Wonderland. For full details read my series **Worship of the Creator Source.**

The Illuminati elite are represented by the Queen of Hearts. If you believe you are a weak, defenceless human being then all the mind control and generations of disinformation have worked. However there is True Reality. By peeking behind the curtain as Dorothy did in the Wizard of Oz we discover this 'reality' is merely a horrible hologrammatic projection into our minds.

At all costs the Demon Worshippers and their Illuminati handlers do not wish for us to discern the Truth. They want us to go round and round on the Wheel of Time.

We are all aspects of the One. Pure Potential in the making. We are beings of energy, immortal and indivisible. We are connected to the One as waves are connected to the ocean.

We cannot be separated from what is *us*. In the beginning there was no beginning and there will be *no* end. We are indivisibly part of the Collective Consciousness that is all there is, was and ever will be at the same moment. Time and space are an illusion.

Life and death are illusions. The Multiverses are all illusions. There is one Consciousness from which all thoughts arise. This is known as Pure Potential. For full details checkout the scientific data on the Unified Field Theory. I hope you are now fully awake and aware. This has been a pleasure.

For full knowledge of the *truth of human history*, what will happen to us in future read: